ENERGY SECTOR STANDARD OF THE PEOPLE' S REPUBLIC OF CHINA

中华人民共和国能源行业标准

Technical Code for Eco-Restoration of Vegetation Concrete on Steep Slope of Hydropower Projects

水电工程陡边坡植被混凝土生态修复技术规范

NB/T 35082-2016

Chief Development Department: China Renewable Energy Engineering Institute

Approval Department: National Energy Administration of the People's Republic of China

Implementation Date: December 1, 2016

China Water & Power Press

中国水利水电出版社

Beijing 2019

图书在版编目（CIP）数据

水电工程陡边坡植被混凝土生态修复技术规范 : NB/T 35082-2016 = Technical Code for Eco-Restoration of Vegetation Concrete on Steep Slope of Hydropower Projects(NB/T 35082-2016) : 英文 / 国家能源局发布. -- 北京 : 中国水利水电出版社, 2019.8(2021.11重印)
ISBN 978-7-5170-7984-2

Ⅰ. ①水… Ⅱ. ①国… Ⅲ. ①水电水利工程－边坡－生态恢复－技术规范－中国－英文 Ⅳ. ①TV223-65

中国版本图书馆CIP数据核字(2019)第190262号

ENERGY SECTOR STANDARD
OF THE PEOPLE'S REPUBLIC OF CHINA
中华人民共和国能源行业标准

Technical Code for Eco-Restoration of Vegetation Concrete on Steep Slope of Hydropower Projects

水电工程陡边坡植被混凝土生态修复技术规范

NB/T 35082-2016

First published 2019
Issued by National Energy Administration of the People's Republic of China
国家能源局　发布
Translation organized by China Renewable Energy Engineering Institute
水电水利规划设计总院　组织翻译
Published by China Water & Power Press
中国水利水电出版社　出版发行

Tel: (+ 86 10) 68545623　68545874
fdy@waterpub.com.cn
Account name: China Water & Power Press
Account number: 0200096319000089691
Address: No.1, Yuyuantan Nanlu, Haidian District, Beijing 100038, China
Organization code: 400014639
http: //www.waterpub.com.cn

中国水利水电出版社微机排版中心
清淞永业（天津）印刷有限公司
184mm×260mm　16 开本　3.25 印张　102 千字
2019 年 8 月第 1 版　2021 年 11 月第 2 次印刷

Price（定价）: ¥ 560.00 (US $ 100.00)

Introduction

This English version is one of China's energy sector standard series in English. Its translation was organized by China Renewable Energy Engineering Institute authorized by National Energy Administration of the People's Republic of China in compliance with relevant procedures and stipulations. This English version was issued by National Energy Administration of the People's Republic of China in Announcement [2018] No.17 dated December 26, 2018.

This version was translated from the Chinese Standard NB/T 35082-2016, *Technical Code for Eco-Restoration of Vegetation Concrete on Steep Slope of Hydropower Projects*, published by China Electric Power Press. In the event of any discrepancy in the implementation, the Chinese version shall prevail.

Many thanks go to the staff from the relevant standard development organizations and those who have provided generous assistance in the translation and review process.

For further improvement of the English version, all comments and suggestions are welcome and should be addressed to:

China Renewable Energy Engineering Institute
No.2 Beixiaojie, Liupukang, Xicheng District, Beijing 100120, China
Website: www.creei.cn

Translating organization:

China Three Gorges University

Translating staff:

ZHOU Mingtao	XI Jing	LI Tianqi	HU Xudong
ZHANG Lin	ZHEN Weiwen	QU Qiong	XUE Cailing
JIN Zhangli			

Review panel members:

CUI Peng	Member of Chinese Academy of Sciences Institute of Mountain Hazards and Environment, CAS
LIU Xiaofen	POWERCHINA Zhongnan Engineering Corporation Limited
QI Wen	POWERCHINA Beijing Engineering Corporation

	Limited
LI Jianlin	China Three Gorges University
HU Xiaoqiong	China Three Gorges University
QIAO Peng	POWERCHINA Northwest Engineering Corporation Limited
HOU Yangyang	POWERCHINA Huadong Engineering Corporation Limited
Altaeb Mohammed	China Three Gorges University
	Wad Madani Technical College
CUI Lei	China Renewable Energy Engineering Institute
HOU Yujing	China Institute of Water Resources and Hydropower Research
WEN Dian	POWERCHINA Chengdu Engineering Corporation Limited
LIU Bin	POWERCHINA Chengdu Engineering Corporation Limited
ZHANG Qian	POWERCHINA Guiyang Engineering Corporation Limited
LI Junyou	China Gezhouba Group No.1 Engineering Corporation Limited
YE Bin	POWERCHINA Huadong Engineering Corporation Limited
LI Qian	POWERCHINA Chengdu Engineering Corporation Limited
Thomas Ramsey	School of Foreign Languages, China Three Gorges University

National Energy Administration of the People's Republic of China

翻译出版说明

本译本为国家能源局委托水电水利规划设计总院按照有关程序和规定，统一组织翻译的能源行业标准英文版系列译本之一。2018 年 12 月 26 日，国家能源局以 2018 年第 17 号公告予以公布。

本译本是根据中国电力出版社出版的《水电工程陡边坡植被混凝土生态修复技术规范》NB/T 35082—2016 翻译的，著作权归国家能源局所有。在使用过程中，如出现异议，以中文版为准。

本译本在翻译和审核过程中，本标准编制单位及编制组有关成员给予了积极协助。

为不断提高本译本的质量，欢迎使用者提出意见和建议，并反馈给水电水利规划设计总院。

地址：北京市西城区六铺炕北小街 2 号

邮编：100120

网址：www.creei.cn

本译本翻译单位：三峡大学

本译本翻译人员：周明涛　席　敬　李天齐　胡旭东　张　琳　甄伟文　屈　琼　薛才玲　金章利

本译本审核人员：

崔　鹏	中国科学院院士 中国科学院水利部成都山地灾害与环境研究所
刘小芬	中国电建集团中南勘测设计研究院有限公司
齐　文	中国电建集团北京勘测设计研究院有限公司
李建林	三峡大学
胡晓琼	三峡大学
乔　鹏	中国电建集团西北勘测设计研究院有限公司
侯洋洋	中国电建集团华东勘测设计研究院有限公司
Altaeb Mohammed	三峡大学

Wad Madani Technical College
崔　磊　水电水利规划设计总院
侯瑜京　中国水利水电科学研究院
文　典　中国电建集团成都勘测设计研究院有限公司
刘　斌　中国电建集团成都勘测设计研究院有限公司
张　倩　中国电建集团贵阳勘测设计研究院有限公司
李俊友　中国葛洲坝集团第一工程有限公司
叶　彬　中国电建集团华东勘测设计研究院有限公司
李　茜　中国电建集团成都勘测设计研究院有限公司
Thomas Ramsey　三峡大学外国语学院

国家能源局

Announcement of National Energy Administration of the People's Republic of China [2016] No.6

According to the requirements of Document GNJKJ [2009] No.52, "Notice on Releasing the Energy Sector Standardization Administration Regulations (*tentative*) and detailed implementation rules issued by National Energy Administration of the People's Republic of China", 144 sector standards such as *Code for Construction Quality Acceptance of Nuclear Power Conventional Island and Balance of Plant Part 8: Thermal Insulation and Painting*, including 75 energy sector (NB) standards, 69 electric power sector (DL) standards, are issued by National Energy Administration of the People's Republic of China after due review and approval.

Attachment: Directory of Sector Standards

National Energy Administration of the People's Republic of China

August 16, 2016

Attachment:

Directory of Sector Standards

Serial number	Standard No.	Title	Replaced standard No.	Adopted international standard No.	Approval date	Implementation date
…						
32	NB/T 35082-2016	Technical Code for Eco-Restoration of Vegetation Concrete on Steep Slope of Hydropower Projects			2016-08-16	2016-12-01
…						

Foreword

According to the requirements of Document GNKJ [2014] No.298 issued by National Energy Administration of the People's Republic of China, "Notice on Releasing the Development and Revision Plan of the First Batch of Energy Sector Standards in 2014", and after extensive investigation and research, summarization of practical experience, consultation of relevant advanced Chinese and foreign standards and wide solicitation of opinions, the drafting group has prepared this code.

The main technical contents of this code include: basic data, drainage, irrigation, reinforcement, vegetation, vegetation concrete, construction, maintenance and management, and inspection.

National Energy Administration of the People's Republic of China is in charge of the administration of this code. China Renewable Energy Engineering Institute has proposed this code and is responsible for its routine management. Energy Sector Standardization Technical Committee on Hydropower Planning, Resettlement and Environmental Protection is responsible for the explanation of specific technical contents. Comments and suggestions in the implementation of this code should be addressed to:

China Renewable Energy Engineering Institute
No.2 Beixiaojie, Liupukang, Xicheng District, Beijing 100120, China

Chief development organizations:

China Three Gorges University

China Renewable Energy Engineering Institute

Participating development organizations:

Management Center of Ministry of Water Resources for Water and Soil Conservation Plant Development

SINOHYDRO Bureau 7 Co., Ltd.

POWERCHINA Kunming Engineering Corporation Limited

China Gezhouba Group Three Gorges Construction Engineering Corporation Limited

China Gezhouba Group Survey Design Corporation Limited

China Gezhouba Group No.6 Engineering Corporation Limited

China Gezhouba Group Foundation Engineering Corporation Limited

Chief drafting staff:

XU Wennian	ZHOU Mingtao	XIA Zhenyao	LIU Daxiang
DING Yu	XU Yang	CUI Lei	ZHAO Bingqin
YANG Yueshu	LI Mingyi	XIA Dong	CHEN Fangqing
WANG Jianzhu	LI Guiyuan	ZHANG Wenli	ZHAO Jiacheng
PEI Dedao	WU Jiangtao	GUO Ping	LI Shaoli
WU Shaoru	SUN Chao	HUANG Xiaole	XI Jing
TAI Yuanlin	CHEN Yuying	JIANG Hao	ZHU Shijiang
SHAN Jie	YANG Xiaotao	WEI Ping	LI Zhengbing
CHEN Pingping	YAO Minghui	WANG Heng	MA Jingchun
SUN Changzhong	JIAO Jiaxun	LI Hantao	HUANG Meng
ZHANG Yuli	LIN Benhua	MAO Yu	AI Lei

Review panel members:

WAN Wengong	YU Weiqi	LI Jianlin	CHEN Shengli
LU Zhaoqin	ZHAO Xinchang	CAO Changbi	ZHANG Xichuan
CHEN Qiuwen	CHANG Jianbo	TAN Shaohua	MAO Yueguang
ZHANG Guodong	ZHOU Yihong	LI Shisheng	

Contents

1 General Provisions

1.0.1 This code is formulated with a view to specifying the design and construction of the vegetation concrete for eco-restoration of steep slopes in hydropower projects.

1.0.2 This code is applicable to stable slopes ranging between 45° and 85° in hydropower projects, and also to similar slopes in other projects.

1.0.3 The eco-restoration vegetation concrete shall be applied to safe and stable slopes. The additional load of vegetation concrete on the slope shall not impair the slope stability.

1.0.4 The eco-restoration works using vegetation concrete shall follow the sustainability principle for long-term service.

1.0.5 The eco-restoration works using vegetation concrete shall be adapted to local conditions, rationally designed, carefully constructed and well maintained and managed with due consideration of such factors as hydrological and meteorological conditions, slope conditions, vegetation conditions, construction conditions and project cost.

1.0.6 Slopes are divided into hard rock slope, soft rock slope, soil-rock slope and barren soil slope. Different types of slopes shall be treated with different methods.

1.0.7 In addition to this code, the eco-restoration of vegetation concrete on steep slopes of hydropower projects shall also comply with the current relevant standards of China.

2 Terms

2.0.1 vegetation concrete

mixture blended with planting soil, cement, organism of habitat material, amendment of habitat material, plant seeds and water, being a typical habitat material with strong scour resistance, high fertility and reasonable distribution of solid-liquid-gas phases

2.0.2 organism of habitat material

granular substance used as a vegetation concrete ingredient which is made by smashing, mixing and fermenting the raw materials of farmyard manure, straw, chaff, sawdust, vinasse and/or other natural organics

2.0.3 amendment of habitat material

fine granular substance used to improve microbial environment, pH value, fertility, water-retaining property, structure and other physical and chemical properties of vegetation concrete

2.0.4 watershed of slope crest

water gathering zone between the intercepting ditch of slope crest and the upper edge of restored slope, which is used to provide water for slope vegetation

3 Basic Data

3.0.1 Before the eco-restoration design, basic data of the project area shall be investigated so as to acquire the information of meteorology, geology, water source, topsoil, natural organic materials, vegetation, etc.

3.0.2 The meteorological investigation shall be in accordance with the following requirements:

1 The investigation contents shall mainly include climate zone and type, mean annual sunshine duration, mean annual temperature, extreme maximum and minimum temperatures, mean annual precipitation, mean annual evaporation, frost-free period, freezing period, wind speed, effective accumulated temperature higher than or equal to 10 °C.

2 The investigation should be mainly based on data collection and analysis with necessary field survey.

3.0.3 The geological investigation data shall be in accordance with the following requirements:

1 The investigation contents shall mainly include slope rock-soil type, slope area, slope aspect, maximum gradient, maximum height, groundwater condition, slope seepage, slope stability and slope pattern.

2 The investigation should be mainly based on data collection and analysis with necessary field survey.

3.0.4 The water source investigation shall be in accordance with the following requirements:

1 The investigation contents shall include tap water, well water and river/lake water, with consideration of water supply capacity, distance, head and cost.

2 The investigation should mainly rely on field survey and be supplemented with data analysis.

3.0.5 The topsoil investigation shall be in accordance with the following requirements:

1 The investigation contents shall mainly include type, texture, structure, availability of topsoil and cost. Priority shall be given to investigating the planting soil in the area to be occupied or disturbed by the eco-restoration project.

2 The investigation should mainly rely on field survey and be

supplemented with data analysis.

3.0.6 The natural organic material investigation shall be in accordance with the following requirements:

1 The investigation contents shall mainly include farmyard manure, straw, chaff, sawdust, vinasse, and the available amount and cost for each type of material.

2 The investigation should mainly rely on field survey and be supplemented with data analysis.

3.0.7 The vegetation investigation shall be in accordance with the following requirements:

1 The investigation contents shall mainly include the regional vegetation types and native plant species around the slope.

2 The investigation should mainly rely on field survey and be supplemented with sampling when necessary.

3.0.8 The investigation contents and record forms shall meet the requirements of Appendix A of this code.

4 Drainage

4.0.1 In the design of interception and drainage, a watershed shall, if conditions allow, be set on the slope crest between the upper edge of the restored slope and the intercepting ditch on the slope crest to replenish water for slope vegetation.

The watershed of slope crest is shown in Figure 4.0.1 below.

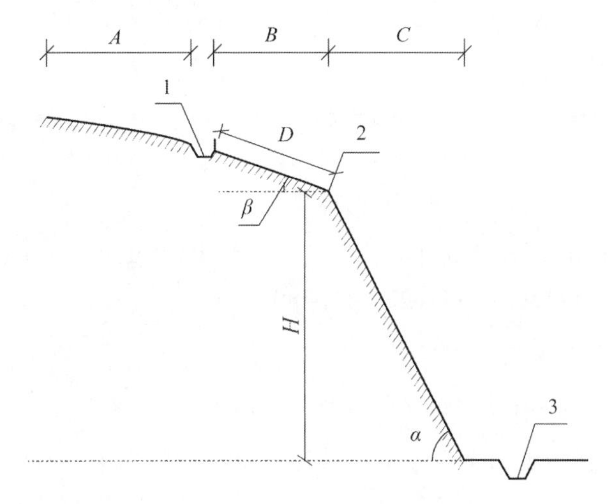

Key

1 intercepting ditch on slope crest

2 upper edge of restored slope

3 drainage ditch at slope toe

α slope angle

β slope angle of watershed

A natural slope

B watershed

C restored slope

H slope height

D distance from the lower edge of intercepting ditch on slope crest to the upper edge of restored slope

Figure 4.0.1 Schematic of watershed of slope crest

4.0.2 When a watershed is set on the slope crest, the distance from the lower edge of intercepting ditch on slope crest to the upper edge of the restored slope should be calculated according to Formula (4.0.2):

$$D=\frac{2.94\eta\psi H^{0.32}}{(\cos\alpha)^{0.06}\cos\beta} \tag{4.0.2}$$

where

D is the distance from the lower edge of intercepting ditch on slope crest to the upper edge of the restored slope (m), taking 5 m if the calculated value is less than 5 m, or 15 m if more than 15 m;

η is the correlation coefficient of the mean annual precipitation, taking 1.15 to 1.10 for the mean annual precipitation 400 mm to 600 mm, 1.10 to 1.05 for 600 mm to 800 mm, 1.05 to 1.00 for 800 mm to 1,200 mm, 1.00 to 0.95 for 1,200 mm to 1,600 mm, 0.95 to 0.90 for 1,600 mm to 2,000 mm, and 0.90 to 0.85 for over 2,000 mm;

ψ is the correlation coefficient of slope vegetation, taking 0.8 for pure herb community, 1.0 for herbaceous-shrub community, and 1.2 for pure shrub community;

H is the slope height (m), if the slope is a multistage one, the vertical height of the highest one shall be taken;

α is the slope angle (°);

β is the slope angle of watershed (°).

4.0.3 The slope drainage design shall comply with the relevant requirements of the current sector standard DL/T 5353, *Design Specification for Slope of Hydropower and Water Conservancy Project.*

4.0.4 The design of slope drainage and intercepting ditches shall meet the relevant requirements of the current national standards GB/T 16453.4, *Comprehensive Control of Soil and Water Conservation—Technical Specification—Small Engineering of Store, Drainage and Draw Water* and GB 51018, *Code for Design of Soil and Water Conservation Engineering.*

5 Irrigation

5.0.1 The eco-restoration vegetation concrete works shall be provided with an irrigation system which should be of the stationary spray or drip type.

5.0.2 The quality of irrigation water shall meet the relevant requirements of the current national standard GB 5084, *Standards for Irrigation Water Quality*. The water shall be subjected to filtration if necessary.

5.0.3 The irrigation should be conducted in conjunction with fertilizing and insecticide spraying.

5.0.4 The material selection and layout of the irrigation system shall meet the relevant requirements of the current national standard GB/T 20203, *Technical Specification for Irrigation Projects with Low Pressure Pipe Conveyance.*

5.0.5 Basic parameters of irrigation shall be in accordance with Table 5.0.5.

Table 5.0.5 Basic parameters of irrigation

No.	Item	Determination method	Index
1	Irrigation intensity	Water collection method	12 mm/h to 18 mm/h
2	Irrigation uniformity	Visual inspection method	≥ 0.85
3	Irrigation coverage	Visual inspection method	≥ 0.98
4	Droplet diameter	Filter paper method	1.0 mm to 3.0 mm

5.0.6 Irrigation shall meet the following requirements:

1 The irrigation time and water amount satisfying the plant growth demand shall be determined in consideration of the natural rainfall and slope surface evaporation.

2 Irrigation shall be conducted in the right amount, multiple times, and in a well-distributed manner.

3 Irrigation under strong sunshine in the summer or early autumn afternoon shall be avoided. In order to prevent plant diseases and pests, irrigation in summer evenings shall be avoided.

6 Reinforcement

6.0.1 Dowels and mesh shall be set for vegetation concrete so as to affix it on the slope surface.

6.0.2 The dowels shall be set to meet the vegetation concrete stability requirements and shall be in accordance with the following requirements:

1 Hot-rolled ribbed rebar should be adopted.

2 The dowel diameter shall not be less than 18 mm.

3 The maximum spacing of dowels on slope surfaces should be calculated according to Formula (6.0.2-1):

$$d = 30[K + \frac{1}{\sin(\alpha - 30^{\circ})}] - 20 \tag{6.0.2-1}$$

where

d is the spacing of dowels (cm);

K is the correlation coefficient of slope type, taking 2.0 for hard rock slopes, 1.5 for soft rock slopes, 1.3 for composite soil-rock slopes and 1.1 for barren soil slopes.

Grades Ⅰ, Ⅱ and Ⅲ slopes specified in the current national standard GB 50218, *Standard for Engineering Classification of Rock Mass*, concrete slopes and masonry slopes are classified as hard rock slopes. Grades Ⅳ and Ⅴ slopes specified in the current national standard GB 50218, *Standard for Engineering Classification of Rock Mass,* are classified as soft rock slopes. The slopes comprising hard rock fragments and sandy or gravelly soils are classified as soil-rock slopes. Clay soil slopes, sandy soil slopes and loess slopes are classified as barren soil slopes.

4 The spacing of dowels at slope perimeter shall be 1/2 of the calculated value of Formula (6.0.2-1).

5 The minimum length of dowels on slope surface should be calculated according to Formula (6.0.2-2):

$$L = \frac{35}{(K - 0.2)^2} + 45\sin(\alpha - 30^{\circ}) + 10 \tag{6.0.2-2}$$

where

L is the minimum length of dowels on slope surface (cm).

6 The length of dowels at slope perimeter shall be increased by 20 cm on the basis of the calculated value of Formula (6.0.2-2).

7 Dowels shall be subjected to corrosion protection.

8 Dowels shall be firmly installed with an exposure length of 8 cm to 12 cm, and the dowels should be angled at 5° to 20°.

6.0.3 Mesh arrangement shall meet the following requirements:

1 Flexible machine-woven wire mesh with a wire diameter not less than 2.0 mm, or flexible plastic mesh with a maximum tensile strength not less than 6.0 kN/m and an ageing resistance not less than 15 years shall be employed.

2 The diameter of mesh cells should be 50 mm to 75 mm.

3 The flexible machine-woven steel wire mesh shall be laminated or galvanized against corrosion.

4 The overlap width of adjacent mesh should be 100 mm to 150 mm.

6.0.4 Mesh shall be firmly bound with dowels or adjacent mesh. The clearance between the mesh and slope surface should be two thirds of the total thickness of sprayed vegetation.

7 Vegetation

7.0.1 Vegetation selection shall meet the following requirements:

1 Native vegetation for slope protection shall be preferably selected, and invasive alien vegetation shall not be used.

2 Vegetation with strong function of adverse resistance, propagation, soil improvement and fixation shall be chosen based on the investigated basic data.

3 Biodiversity and sustainability shall be observed.

4 Better ornamental vegetation may be chosen according to landscape requirements.

7.0.2 Plant seeds and seedlings shall meet the following requirements:

1 Plant seeds shall be indicated with strain, origin, producer, harvest year, purity, germination percentage and thousand-grain weight.

2 Seedlings shall have well developed root systems, healthy and strong stems, and shall be free of injury, pollution in stems or leaves, diseases and pests.

3 Purchased plant seeds and seedlings shall be qualified with domestic inspection and quarantine certificates.

7.0.3 Vegetation arrangement shall adhere to the following principles:

1 Herbage, shrub and vine should be properly combined to coordinate with the surrounding environment. Herbage should dominate for hard and soft rock slopes, and shrub for composite soil-rock slopes and barren soil slopes.

2 Cool-season and warm-season vegetation should be combined according to the investigated basic data.

3 Mutually exclusive vegetation species shall not be planted in the same area.

7.0.4 Pretreatment of plant seeds and seedlings shall meet the following requirements:

1 The purity, germination percentage and thousand-grain weight of plant seeds shall be checked.

2 Plant seeds shall be disinfected and soaked, even shell broken if necessary.

3 Before transplantation, non-potted seedlings shall be temporarily transplanted according to the current sector standard CJJ 82, *Code for Construction and Acceptance of Landscaping Engineering.*

7.0.5 The sowing quantity of plant seeds should be calculated according to Formulae (7.0.5-1) and (7.0.5-2):

$$A=\sum k_i A_i \tag{7.0.5-1}$$

$$A_i=\frac{N_i Z_i}{(1-R_i)\,C_i\,F_i}\times 10^{-3} \tag{7.0.5-2}$$

where

A is the total sowing quantity (g/m^2);

A_i is the monoculture sowing quantity (g/m^2);

k_i is the polyculture sowing ratio (%), $\sum k_i=1$, which is determined by the vegetation configuration plan and requirements, climatic conditions, slope characteristics, landscape requirements, etc.;

N_i is the number of sowing seeds per unit area in the case of monoculture ($grains/m^2$);

Z_i is the thousand-grain weight of sole-plant (g);

R_i is the spraying loss rate of sole-plant seeds (%), taking 5 % for thousand-grain weight less than 0.5 g, 10 % for 0.5 g to 1.0 g, 15 % for 1.0 g to 5.0 g, and 20 % for over 5.0 g;

C_i is the purity of sole-plant seeds (%);

F_i is the germination percentage of sole-plant seeds (%).

C_i and F_i test methods shall comply with the relevant requirements of the current national standard GB 2772, *Rules for Forest Tree Seed Testing.*

8 Vegetation Concrete

8.1 Materials

8.1.1 Vegetation concrete shall be made by mixing planting soil, organism of habitat material, cement, amendment of habitat material, plant seeds and water as per the required proportion.

8.1.2 Planting soil shall be in accordance with the following requirements:

1 An appropriate topsoil source shall be selected according to the topsoil investigation.

2 The main physical and chemical indexes of the topsoil samples after being dried, crushed and sieved, should be in accordance with Table 8.1.2.

Table 8.1.2 Main physical and chemical indexes of planting soil

No.	Item	Index
1	Total Cd	≤ 1.5 mg/kg
2	Total Hg	≤ 1.0 mg/kg
3	Total Pb	$\leq 1.2\times10^2$ mg/kg
4	Total Cr	≤ 70 mg/kg
5	Total As	≤ 10 mg/kg
6	Total Ni	≤ 60 mg/kg
7	Total Zn	$\leq 3.0\times10^2$ mg/kg
8	Total Cu	$\leq 1.5\times10^2$ mg/kg
9	pH	5.5 to 8.5
10	Salt content	≤ 1.5 g/kg
11	Total nutrient	≥ 0.20 %
12	Particle size	≤ 8.0 mm
13	Moisture content	≤ 20 %

NOTES:

1 The indexes of heavy metals are calculated based on the mass of oven-dried soil.

2 The test methods of heavy metals, pH, salt content and total nutrient shall comply with the relevant requirements of the current sector standard HJ/T 166, *The Technical Specification for Soil Environmental Monitoring*. The particle size shall be tested by the pipette method, and moisture content by the oven-drying method.

3 Total nutrient = Total Nitrogen (N) + Total Phosphorus (P_2O_5) + Total Potassium (K_2O).

8.1.3 Organisms of habitat materials shall meet the following requirements:

1 Several kinds of raw materials should be selected according to the investigation of natural organic materials.

2 The main indexes of the sampled natural organic materials after being smashed, mixed and fermented should be in accordance with those specified in Table 8.1.3.

Table 8.1.3 Main indexes of organism of habitat materials

No.	Item	Index
1	Particle size	≤ 8.0 mm
2	pH	5.5 to 8.5
3	EC	0.50 mS/cm to 3.0 mS/cm
4	Total nutrient	≥ 1.5 %
5	Moisture content	≤ 20 %
6	Aeration porosity	≥ 15 %

NOTE The test method of each index in the table shall comply with the relevant requirements of the current sector standard LY/T 1970, *Organic Media for Greening Use*.

8.1.4 Portland cement P.O 42.5 should be chosen, and its main indexes shall meet the relevant requirements of the current national standard GB 175, *Common Portland Cement*.

8.1.5 The main indexes of amendment of habitat material shall be in accordance with those specified in Table 8.1.5.

Table 8.1.5 Main indexes of amendment of habitat materials

No.	Item	Index
1	Specific surface area	$\geq 1.5\times10^{2}$ m^2/kg
2	Moisture holding capacity	13 kg/cm^2 to 15 kg/cm^2
3	Quantity of microorganism	1.0×10^{8} CFU/g to 1.0×10^{9} CFU/g
4	Total nutrient	≥ 8.5 %
5	pH	≤ 4.5

NOTE The specific surface area shall be tested by the Blaine method, and the test method of water holding capacity, Quantity of microorganism, total nutrient and pH shall comply with the relevant requirements of the current sector standard HJ/T 166, *The Technical Specification for Soil Environmental Monitoring*.

8.1.6 Water quality shall meet the relevant requirements of the current national standard GB 5084, *Standards for Irrigation Water Quality*.

8.2 Proportioning

8.2.1 Vegetation concrete comprises a base layer and a surface layer, both of which shall be prepared separately. In the preparation of the base layer, solid-phase mixture is composed of planting soil, organism of habitat materials, cement, amendment of habitat material. In the preparation of the surface layer, plant seeds are added into the above-mentioned mixture.

8.2.2 With the volume of planting soil as the benchmark, the amount of organism of habitat material, cement, amendment of habitat material shall meet the following requirements:

1 The volume of organism of habitat material should be calculated according to Formula (8.2.2-1):

$$V_{om} = (0.25 + 0.35 K_a K \frac{\alpha - 45^\circ}{90^\circ}) V_{ps} \tag{8.2.2-1}$$

where

V_{om} is the volume of organism of habitat materials (m^3);

K_a is the correlation coefficient of climate zone, taken as shown in Table 8.2.2;

V_{ps} is the volume of planting soil (m^3).

Table 8.2.2 Climate zone correlation coefficient K_a

Climatic zone	Climatic region	
	Humid region A	Sub-humid region B
Mid-temperate	1.05	1.10
Warm-temperate	1.00	1.05
North subtropical	1.00	–
Mid-subtropical	1.00	–
South subtropical	0.950	1.05
Marginal tropical	0.900	0.950
Mid-tropical	0.900	–
Subtropical plateau mountain region	0.950	
Temperate plateau	1.05	1.10

2 The mass of cement should be calculated according to Formula (8.2.2-2):

$$M_c = K_l(0.06 + 0.07\frac{K}{K_\alpha}\frac{\alpha - 45^\circ}{90^\circ})V_{ps}\rho_{ps} \tag{8.2.2-2}$$

where

M_c is the mass of cement (kg);

K_l is the correlation coefficient of the base layer or surface layer, taking 1.0 for the base layer and 0.5 for the surface layer;

ρ_{ps} is the dry density of the planting soil (kg/m^3).

3 The mass of amendment of habitat materials should be calculated according to Formula (8.2.2-3):

$$M_{aa} = 0.5M_c \tag{8.2.2-3}$$

where

M_{aa} is the mass of amendment of the habitat materials (kg).

8.2.3 In the process of mixing, the solid-phase mixture shall be in accordance with the following requirements:

1 Mixing shall be conducted evenly and mechanically at or near the construction site.

2 Materials shall be fed in sequence: firstly planting soil; secondly organism of habitat materials, cement and amendment of habitat materials; and finally plant seeds.

3 The single mixing time shall be conducted for 3 min to 5 min.

8.2.4 The water amount shall be proper so that the vegetation concrete sprayed on the slope surface would not scatter or flow.

8.3 Spraying

8.3.1 The spraying angle of spray gun shall be no more than 15°, and the distance between spray gun nozzle and slope surface should be 0.8 m to 1.2 m.

8.3.2 Spraying shall be conducted in two steps, the base layer first and then the surface layer.

8.3.3 The ready-made solid-phase mixture shall be sprayed within 6 hours.

8.3.4 The spraying thickness of the base layer shall be as shown in Table 8.3.4, and that of the surface layer should be 20 mm.

Table 8.3.4 Spraying thickness of base layer

Slope type	Mean annual precipitation (mm)	Slope angle	Thickness (mm)
Hard rock slope	≤900	70° to 85°	90
		45° to 70°	100
	> 900	70° to 85°	80
		45° to 70°	90
Soft rock slopc	≤900	65° to 85°	80
		45° to 65°	90
	> 900	65° to 85°	70
		45° to 65°	80
Composite soil-rock slope	≤900	65° to 85°	60
		45° to 65°	70
	> 900	65° to 85°	50
		45° to 65°	60
Barren soil slope	≤600	45° to 85°	50
	600 to 1,200		40
	≥1,200		30

8.3.5 The time interval between the base layer spraying and the surface layer spraying shall be less than 4 hours.

8.3.6 Spraying shall be conducted evenly to avoid missing, especially for irregularities or corner areas of slope surface.

8.3.7 Spraying should not be conducted when wind speed is greater than 10.8 m/s or when it rains.

8.3.8 Vegetation concrete spraying shall be tested timely, and the inspection indexes shall be in accordance with those specified in Table 8.3.8. If the inspection result is not qualified, the mix proportion shall be promptly adjusted.

Table 8.3.8 Inspection indexes of vegetation concrete

No.	Item	Test method	Index		
			1 d	3 d	≥ 28 d
1	Bulk density	Ring sampler	1.3 g/cm^3 to 1.7 g/cm^3		
2	Aeration porosity	Ring sampler	≥ 25 %		
3	pH	Potentiometric method	6.0 to 8.5		
4	Moisture content	Oven-drying method	≥ 15 %		
5	Hydrolyzable nitrogen	Alkali N-proliferation	≥ 60 mg/kg		
6	Available phosphorus	Mo-Sb colorimetric method	≥ 20 mg/kg		
7	Available potassium	Flame photometry	≥ 1.0×10^2 mg/kg		
8	Restored degree of shrinkage	Ring sampler	≥ 90 %		
9	Quantity of microorganism	The methods specified in the current sector standard HJ/T 166, The *Technical Specification for Soil Environmental Monitoring*	≥ 1.0×10^6 CFU/g		
10	Unconfined compressive strength (MPa)	The methods specified in the current national standard GB/T 50123, *Standard for Soil Test Method*	0.25 to 0.45	0.40 to 0.55	≥ 0.38
11	Erosion modulus (80 mm/h rainfall intensity)	The methods specified in the relevant requirements of the current sector standard SL 419, *Test Specification of Soil and Water Conservation*	≤ 3.0×10^2 g/(m^2·h)	≤ 1.0×10^2 g/(m^2·h)	

9 Construction

9.1 Construction Preparation

9.1.1 Construction shall be carried out according to the relevant documents of slope works and eco-restoration technology of vegetation concrete. The site administrators and constructors shall be familiar with design purposes and requirements before construction.

9.1.2 A construction planning shall be prepared before the construction, including the following contents:

1 Construction conditions.

2 Construction procedures and methods.

3 Construction layouts.

4 Resources allocation.

5 Construction quality assurance plan.

6 Construction safety assurance and environmental protection plan.

7 Construction schedule.

9.1.3 The construction materials, equipment and facilities shall be prepared according to the construction schedule.

9.1.4 Materials on site shall be protected against water, sunshine and corrosion.

9.1.5 Spraying shall be executed in seasons favorable for plant seeds germination.

9.2 Construction Procedure

9.2.1 Construction shall be carried out after the acceptance of the works such as slope excavation, slope reinforcement and underground pipeline laying.

9.2.2 Construction process should comply with the requirements in Figure 9.2.2.

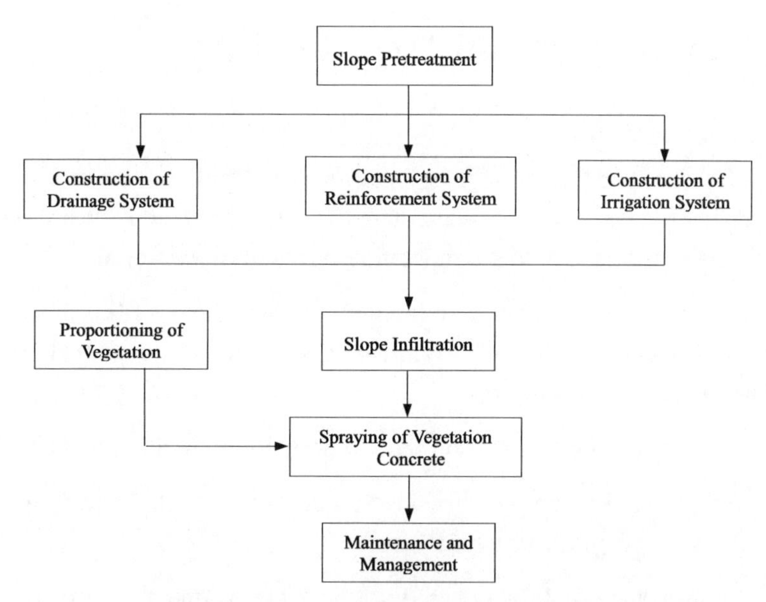

Figure 9.2.2 Construction process

9.2.3 Loose materials, such as unsteady rocks, loose soil and exposed roots shall be cleared in the process of slope pretreatment. The reverse slopes or sagging slopes should be treated by slope cutting or mortared masonry.

9.2.4 Construction of drainage system shall include a series of works, such as the intercepting ditch on slope crest, the drainage ditch at slope toe and the water seepage treatment of slope surface.

9.2.5 The reinforcement system shall be constructed in the sequence of laying mesh, installing dowels, and binding the dowels with the mesh.

9.2.6 Construction of irrigation system shall include a series of works, such as pipeline laying, sprinkler installation and irrigation water filtration.

9.2.7 Infiltration shall be conducted to moisturize the slope, and the infiltration duration shall not be less than 48 hours.

9.2.8 Spraying shall be executed within 3 hours after the slope infiltration.

9.2.9 The acceptance forms of construction works shall meet the requirements of Appendix B of this code.

10 Maintenance and Management

10.1 Seedling Stage

10.1.1 After sprayed, the vegetation concrete shall be maintained and managed for 60 days as the seedling stage, and the duration may be extended appropriately in the case of low temperature or insufficient rainfall.

10.1.2 The maintenance and management during the seedling stage shall include slope surface mulching, irrigation, disease and pest control, seedling replanting, local defect repair, etc.

10.1.3 Slope surface mulching shall meet the following requirements:

1 Mulch may bc non-woven fabrics, sunshade net, etc., and in winter, it also may be straw, straw mat, etc.

2 Slope surface shall be covered within 2 hours after the surface layer is sprayed.

3 Mulch shall be laid firmly and in close contact with the slope surface.

4 Plastic membrane shall be applied on the slope surface if heavy rainfall occurs within 4 hours after spraying work.

10.1.4 Slopes sprayed with vegetation concrete shall be inspected once every day, and inspection contents shall include slope vegetation moisture, plant seed germination or seedling survival, disease and pest, vegetation concrete stability, etc.

10.1.5 Disease and pest control shall meet the following requirements:

1 The inspection of disease and pest shall be enhanced, and the prevention and control measures shall be taken timely if any problem is found.

2 The disease and pest shall be controlled by corresponding biological, physical and/or chemical means according to the real situation.

3 The chemicals used for disease and pest control shall have high efficiency and low toxicity, and be safe to the natural predators. The application of the agrochemicals shall be in strict accordance with the instructions.

10.1.6 Replanting shall be carried out promptly in the case of seedling death.

10.1.7 Local defect repair shall be in accordance with the following requirements:

1 When bare spots or spallings are found on vegetation concrete, the

causes shall be identified to eliminate hidden dangers, and the defects shall be repaired timely.

2 When the local defect area is small, manual resowing or seedling transplanting may be adopted.

3 When the local defect area is large, loose materials on the corresponding spots shall be removed first, and vegetation concrete shall be re-sprayed.

10.2 Growth Stage

10.2.1 After the seedling stage maintenance and management, the vegetation concrete shall be maintained and managed for no less than 240 days for seedlings growth, and the duration may be extended appropriately under special circumstances.

10.2.2 The maintenance and management during the growth stage shall include irrigation, disease and pest control, seedling replanting, local defect repair, etc.

10.2.3 Slopes sprayed with vegetation concrete shall be inspected once every two weeks, and inspection contents shall include slope vegetation moisture, plant growth, disease and pest, vegetation concrete stability, etc.

10.2.4 Disease and pest control shall meet the requirements of the Article 10.1.5 of this code.

10.2.5 Local defect repair shall meet the requirements of the Article 10.1.7 of this code.

11 Inspection

11.1 Material

11.1.1 Factory certificates and product quality certificates shall be checked when purchasing the materials such as cement, amendment of habitat materials, mesh, dowels, irrigation pipes. When purchasing plant seeds and seedlings, the inspection and quarantine certificates shall be checked.

11.1.2 Before using the materials on site, including cement, amendment of habitat materials, mesh, dowels, irrigation pipes, planting soil, organism of habitat material, plant seeds, seedlings and water, random sampling and inspection for each batch shall be conducted, and the inspection reports shall be prepared. The formats and contents of inspection reports shall comply with Appendix C of this code.

11.1.3 The quantity of material to form an inspection lot shall be as below:

1 The quantity of each inspection lot is 20 tons for cement, 10 tons for amendment of habitat materials, 2,000 m^2 for mesh, 2,000 pieces for dowels; 200 m^3 for planting soil, 60 m^3 for organism of habitat material, one pack for plant seeds, and 500 plants for seedlings. At least one inspection lot shall be taken for each water source. One inspection lot shall be taken for each batch of supplied materials or discontinuously supplied materials, even less than the quantity of one inspection lot specified above.

2 The quantity of each inspection lot may be doubled in one of the following cases:

1) Materials have been certificated and approved by product certification authorities.

2) Materials with reliable sources and having passed the inspection for three consecutive times.

3) Materials from the same batch and the same factory, which are used in several works of the same project.

11.1.4 Three samples shall be taken from each batch of materials for inspection.

11.1.5 The inspection data shall be processed as below:

1 Among the inspection values of three samples, if the difference between the maximum value and the median value, and the difference between the median value and the minimum value are both no more than 15 % of the median value, the arithmetic mean value of the 3 inspection values shall be taken.

2 Among the inspection values of three samples, if either the difference between the maximum value and the median value or the difference between the median value and the minimum value is more than 15 % of the median value, the median value shall be taken.

3 Among the inspection values of three samples, if the differences between the maximum value and the median value, and difference between the median value and the minimum value are both more than 15 % of the median value, the material shall not be used.

11.2 Vegetation Concrete

11.2.1 Each 1,000 m^2 slope surface sprayed with vegetation concrete mixed with the same materials and the same mix proportion shall be inspected at least once with three samples.

11.2.2 The inspection items, methods and indexes for vegetation concrete performance shall meet the requirements of Article 8.3.8 of this code.

11.2.3 The inspection data shall be processed as specified in Article 11.1.5 of this code.

11.2.4 The inspection contents and record formats shall meet the requirements of Appendix D of this code.

Appendix A Investigation Contents and Record Formats of Basic Data

A.0.1 The investigation contents and record formats of meteorological data shall comply with Table A.0.1.

Table A.0.1 Investigation contents and record formats of meteorological data

Project Name		**Project Location**		
No.	**Item**	**Unit**	**Result**	**Remarks**
1	Climate zone			
2	Climate type			
3	Mean annual sunshine hours	h		
4	Mean annual temperature	°C		
5	Extreme maximum temperature	°C		
6	Extreme minimum temperature	°C		
7	Mean annual precipitation	mm		
8	Mean annual evaporation	mm		
9	Frost-free period	d		
10	Freezing period	d		
11	Effective accumulated temperature higher than or equal to 10 °C	°C		
12	Days and distribution with a wind speed not less than 10.8 m/s:			
Investigator	Signature:		Date:	

NOTES:

1 Climate zones include the mid-temperate zone, warm-temperate zone, north subtropical zone, mid-subtropical zone, south subtropical zone, marginal tropical zone, mid-tropical zone, subtropical plateau mountain region, plateau temperate zone, etc.

2 Climate types include polar climate, temperate continental climate, temperate marine climate, monsoon climate of medium latitudes, subtropical monsoon climate, tropical desert climate, tropical savanna climate, tropical rainforest climate, tropical monsoon climate, mediterranean climate and alpine plateau climate.

A.0.2 The investigation contents and record formats of geological data shall comply with Table A.0.2.

Table A.0.2 Investigation contents and record formats of geological data

Project name		Project location		
No.	**Item**	**Unit**	**Result**	**Remarks**
1	Type of rock-soil slope			
2	Slope area	m^2		
3	Slope aspect			
4	Maximum slope angle	(°)		
5	Maximum slope height	m		
6	Groundwater condition			
7	Slope seepage condition			
8	Slope stability			
9	Slope pattern description including reverse slope, surface evenness and slope grade, etc.			
Investigator	Signature:		Date:	

NOTES:

1 Type of rock-soil slope refers to the hard rock slope, soft rock slope, soil-rock slope or barren soil slope.

2 Slope aspects include sunny slope and shaded slope.

3 Slope stability refers to being stable or unstable.

A.0.3 The investigation contents and record formats of water source data shall comply with those specified in Table A.0.3.

Table A.0.3 Investigation contents and record formats of water source data

Project name			Project location	
Item		**Unit**	**Result**	**Remarks**
Tap water	Supply capacity	m^3/d		
	Distance	m		
	Head	m		
	Cost	CNY/m^3		

Table A.0.3 *(continued)*

Project name			**Project location**	
Item		**Unit**	**Result**	**Remarks**
Well water	Supply capacity	m^3/d		
	Distance	m		
	Head	m		
	Cost	CNY/m^3		
River/lake water	Supply capacity	m^3/d		
	Distance	m		
	Head	m		
	Cost	CNY/m^3		
Investigator	Signature:			Date:

NOTES:

1 The investigation focuses on tap water, well water and river/lake water.
2 Distance refers to the length of the path from water source to slope location.
3 Head refers to the difference in height between water intake and slope crest.
4 Cost is measured by the cost of water up to the project site.

A.0.4 The investigation contents and record formats of topsoil source data shall comply with those specified in Table A.0.4.

Table A.0.4 Investigation contents and record formats of topsoil source data

Project name		**Project location**	
No.	**Item**	**Result**	
1	Soil type		
2	Soil texture		
3	Soil structure		
4	Cost		
5	Available amount of topsoil		
Investigator	Signature:		Date:

NOTES:

1 Soil types include red earth, yellow earth, brown earth, cinnamon soil, calcium soil, black loam soil, desert soil, alpine meadow soil, and alpine desert soil.
2 Soil texture refers to sandy soil, loam or clay.
3 Soil structure refers to be granular, massive, columnar, or schistose, etc.
4 Cost is measured by the cost of soil up to the project site.

A.0.5 The investigation contents and record formats of natural organic materials data shall comply with those specified in Table A.0.5.

Table A.0.5 Investigation contents and record formats of natural organic materials data

Project name			**Project location**	
Item		**Unit**	**Result**	**Remarks**
Farmland manure	Available amount	m^3		
	Cost	CNY/ m^3		
Straw	Available amount	m^3		
	Cost	CNY/ m^3		
Chaff	Available amount	m^3		
	Cost	CNY/ m^3		
Sawdust	Available amount	m^3		
	Cost	CNY/ m^3		
Vinasse	Available amount	m^3		
	Cost	CNY/ m^3		
Investigator	Signature:			Date:

NOTES:

1 The investigation focuses on five types of natural organic materials: farmland manure, straw, chaff, sawdust and vinasse.

2 Available amount refers to the quantity which can be obtained within 30 km near the project location.

3 Cost is measured by the cost of natural organic materials up to the project site.

A.0.6 The investigation contents and record formats of vegetation data shall comply with those specified in Table A.0.6.

Table A.0.6 Investigation contents and record formats of vegetation data

Project name		**Project location**	
No.	**Item**	**Result**	
1	Regional vegetation type		

Table A.0.6 *(continued)*

Project name				**Project location**	
No.	**Item**			**Result**	
2	Native plant species	Sunny slope	Herbage		
			Shrub		
			Arbor		
			Flower		
			Vine		
		Shaded slope	Herbage		
			Shrub		
			Arbor		
			Flower		
			Vine		
Investigator	Signature:			Date:	

NOTES:

1 Regional vegetation types refer to the herb-dominated vegetation, shrub-dominated vegetation, herbaceous-shrub vegetation, arbor-shrub vegetation, with the description of the natural matching and growing conditions of arbor, shrub, herbage near the slope.

2 Native plant species include the main varieties of herbage, shrub, arbor, flower and vine.

Appendix B Acceptance Forms of Construction Works*

B.0.1 The acceptance of slope pretreatment works shall be in accordance with those specified in Table. B.0.1.

Table B.0.1 Acceptance form of slope pretreatment works

<table>
<tr><td>Project name</td><td colspan="2"></td><td>Project location</td><td></td></tr>
<tr><td colspan="3">Item</td><td colspan="2">Result</td></tr>
<tr><td colspan="3">Stability of slope base layer</td><td colspan="2"></td></tr>
<tr><td colspan="3">Treatment of loose materials on slope</td><td colspan="2"></td></tr>
<tr><td colspan="3">Trimming of reverse slope</td><td colspan="2"></td></tr>
<tr><td colspan="3">Treatment of concave slope</td><td colspan="2"></td></tr>
<tr><td colspan="3">Removal of spoil at slope toe</td><td colspan="2"></td></tr>
<tr><td colspan="3">Slope drainage</td><td colspan="2"></td></tr>
<tr><td colspan="3">Slope infiltration</td><td colspan="2"></td></tr>
<tr><td colspan="3">Treatment of slope seepage</td><td colspan="2"></td></tr>
<tr><td colspan="3">Others</td><td colspan="2"></td></tr>
<tr><td rowspan="3" colspan="2">Acceptance comments</td><td>Contractor</td><td>Signature:</td><td>Date:</td></tr>
<tr><td>Supervisor</td><td>Signature:</td><td>Date:</td></tr>
<tr><td>Employer</td><td>Signature:</td><td>Date:</td></tr>
</table>

* *Translator's Annotation: The "Acceptance Forms of Construction Works" previously written as the "Receiving reports of construction processes" in the Chinese Version has been corrected in this English Version.*

B.0.2 The acceptance of reinforcement works shall be in accordance with those specified in Table B.0.2.

Table B.0.2 Acceptance form of reinforcement works

<table>
<tr><td>Project name</td><td colspan="2"></td><td>Project location</td><td></td></tr>
<tr><td colspan="3">Item</td><td colspan="2">Result</td></tr>
<tr><td colspan="3">Dowel material (ribbed or round)</td><td colspan="2"></td></tr>
<tr><td colspan="3">Dowel spacing (slope surface and slope perimeter)</td><td colspan="2"></td></tr>
<tr><td colspan="3">Dowel nominal diameter</td><td colspan="2"></td></tr>
<tr><td colspan="3">Dowel length</td><td colspan="2"></td></tr>
<tr><td colspan="3">Grouting condition of anchor hole</td><td colspan="2"></td></tr>
<tr><td colspan="3">Exposure length of dowel</td><td colspan="2"></td></tr>
<tr><td colspan="3">Corrosion protection of dowel</td><td colspan="2"></td></tr>
<tr><td colspan="3">Anchor angle</td><td colspan="2"></td></tr>
<tr><td colspan="3">Mesh type</td><td colspan="2"></td></tr>
<tr><td colspan="3">Wire diameter of flexible machine-woven wire mesh</td><td colspan="2"></td></tr>
<tr><td colspan="3">Mesh diameter</td><td colspan="2"></td></tr>
<tr><td colspan="3">Maximum stretching force of flexible plastic mesh</td><td colspan="2"></td></tr>
<tr><td colspan="3">Ageing resistance of flexible plastic mesh</td><td colspan="2"></td></tr>
<tr><td colspan="3">Overlap situation of mesh</td><td colspan="2"></td></tr>
<tr><td colspan="3">Binding situation of mesh (mesh- mesh, mesh-dowel)</td><td colspan="2"></td></tr>
<tr><td rowspan="3">Acceptance comments</td><td>Contractor</td><td colspan="2">Signature:</td><td>Date:</td></tr>
<tr><td>Supervisor</td><td colspan="2">Signature:</td><td>Date:</td></tr>
<tr><td>Employer</td><td colspan="2">Signature:</td><td>Date:</td></tr>
</table>

B.0.3 The acceptance of vegetation concrete spraying works shall be in accordance with those specified in Table B.0.3.

Table B.0.3 Acceptance form of vegetation concrete spraying works

<table>
<tr><td>Project name</td><td colspan="2"></td><td>Project location</td><td></td></tr>
<tr><td colspan="3">Item</td><td colspan="2">Result</td></tr>
<tr><td colspan="3">Mix proportion</td><td colspan="2"></td></tr>
<tr><td colspan="3">Feeding sequence</td><td colspan="2"></td></tr>
<tr><td colspan="3">Mixing mode</td><td colspan="2"></td></tr>
<tr><td colspan="3">Mixing duration</td><td colspan="2"></td></tr>
<tr><td colspan="3">Storage time of mixture</td><td colspan="2"></td></tr>
<tr><td colspan="3">Thickness of base layer</td><td colspan="2"></td></tr>
<tr><td colspan="3">Thickness of surface layer</td><td colspan="2"></td></tr>
<tr><td colspan="3">Evenness of spraying</td><td colspan="2"></td></tr>
<tr><td colspan="3">Water consumption of spraying</td><td colspan="2"></td></tr>
<tr><td colspan="3">Time interval between the base layer spraying and the surface layer spraying</td><td colspan="2"></td></tr>
<tr><td rowspan="3">Acceptance comments</td><td>Contractor</td><td colspan="2">Signature:</td><td>Date:</td></tr>
<tr><td>Supervisor</td><td colspan="2">Signature:</td><td>Date:</td></tr>
<tr><td>Employer</td><td colspan="2">Signature:</td><td>Date:</td></tr>
</table>

Appendix C Inspection Contents and Record Formats of Main Materials

C.0.1 The inspection contents and record formats of planting soil shall be in accordance with those specified in Table C.0.1.

Table C.0.1 Inspection contents and record formats of planting soil*

Project name				Project location	
No.	**Item**	**Method**	**Result**	**Index requirement**	**Remarks**
1	Maximum particle size			≤ 8.0 mm	
2	Moisture content			≤ 20 %	
3	Salt content			≤ 1.5 g/kg	
4	pH			5.5 to 8.5	
5	Total Cd			≤ 1.5 mg/kg	
6	Total Hg			≤ 1.0 mg/kg	
7	Total Pb			≤ 1.2×10^2 mg/kg	
8	Total Cr			≤ 70 mg/kg	
9	Total As			≤ 10 mg/kg	
10	Total Ni			≤ 60 mg/kg	
11	Total Zn			≤ 3.0×10^2 mg/kg	
12	Total Cu			≤ 1.5×10^2 mg/kg	
13	Total nutrient			≥ 0.20 %	
Inspector		Signature:		Date:	

* *Translator's Annotation: The units of No. 6 to No. 12 in Form C.0.1 have been corrected to mg/kg in the English version from g/kg in the Chinese version.*

C.0.2 The inspection contents and record formats of organism of habitat material shall be in accordance with those specified in Table C.0.2.

Table C.0.2 Inspection contents and record formats of organism of habitat material

Project name				Project location	
No.	**Item**	**Method**	**Result**	**Index requirement**	**Remarks**
1	Particle size			≤ 8.0 mm	
2	Impurity content			≤ 5.0 %	
3	pH			5.5 to 8.5	
4	EC			0.50 mS/cm to 3.0 mS/cm	
5	Total nutrient			≥ 1.5 %	
6	Moisture content			≤ 20 %	
7	Aeration porosity			≥ 15 %	
Inspector		Signature:		Date:	

C.0.3 The inspection contents and record formats of cement shall be in accordance with those specified in Table C.0.3.

Table C.0.3 Inspection contents and record formats of cement

Project name			Project location	
No.	**Item**	**Result**	**Index requirement**	**Remarks**
1	Variety		Ordinary Portland cement	
2	Strength grade		P.O 42.5	
3	Production date			
4	Validity period			
5	Quality certificate			
Inspector		Signature:	Date:	

C.0.4 The inspection contents and record formats of amendment of habitat material shall be in accordance with those specified in Table C.0.4.

Table C.0.4 Inspection contents and record formats of amendment of habitat material

Project name				Project location	
No.	Item	Method	Result	Index requirement	Remarks
1	Specific surface area			$\geq 1.5\times10^2$ m^2/kg	
2	Moisture holding capacity			13 kg/cm^2 to 15 kg/cm^2	
3	Quantity of microorganism			1.0×10^8 CFU/g to 1.0×10^9 CFU/g	
4	Total nutrient			≥ 8.5 %	
5	pH			≤ 4.5	
6	Production date				
7	Validity period				
8	Quality certificate				
Inspector		Signature:			Date:

C.0.5 The inspection contents and record formats of vegetation shall be in accordance with those specified in Table C.0.5.

Table C.0.5 Inspection contents and record formats of vegetation

Project name			Project location		
Item	Plant seeds					
Harvest year						
Purity (%)						
Germination percentage (%)						
Thousand-gra in weight(g)						

Table C.0.5 *(continued)*

Project name			**Project location**		
Item	**Plant seeds**				
Origin					
Producer					
Description of vegetation seedling					
Inspector	Signature:		Date:		

NOTES:

1 The inspection of vegetation includes that of seeds and seedlings.

2 The inspection of seedlings includes root system, plant type, diseases and pests, etc.

C.0.6 The inspection contents and record formats of water quality shall be in accordance with those specified in Table C.0.6.

Table C.0.6 Inspection contents and record formats of water quality

Project name				**Project location**	
No.	**Item**	**Method**	**Result**	**Index requirement**	**Remarks**
1	BOD_5			$\leq 1.0\times10^2$ mg/L	
2	COD			$\leq 2.0\times10^2$ mg/L	
3	Suspended substance			$\leq 1.0\times10^2$ mg/L	
4	Anionic surfactant			≤ 8.0 mg/L	
5	Water temperature			$\leq 25\,^\circ C$	
6	Total salt content			$\leq 1.0\times10^3$ mg/L	
7	Chloride			$\leq 3.5\times10^2$ mg/L	
8	Sulfide			≤ 1.0 mg/L	
9	Total Hg			$\leq 1.0\times10^{-2}$ mg/L	
10	Cd			≤ 0.10 mg/L	

Table C.0.6 *(continued)*

Project name				Project location	
No.	**Item**	**Method**	**Result**	**Index requirement**	**Remarks**
11	Total As			≤ 0.10 mg/L	
12	Chromium (hexad)			≤ 0.10 mg/L	
13	Pb			≤ 0.20 mg/L	
14	Fecal coliform bacteria			≤ 4.0×10^3 /100 mL	
15	Ascaris eggs			≤ 2.0 /L	
16	pH			5.5 to 8.5	
Inspector		Signature:		Date:	

C.0.7 The inspection contents and record formats of reinforcement materials shall be in accordance with those specified in Table C.0.7.

Table C.0.7 Inspection contents and record formats of reinforcement materials

Project name		Project location		
Item		**Result**	**Index requirement**	**Remarks**
Dowel	Surface shape		Ribbed	
	Production process		Hot-rolling	
	Diameter			
	Length			
	Corrosion protection			
	Yield strength			

Table C.0.7 *(continued)*

<table>
<tr><td colspan="2">Project name</td><td colspan="2"></td><td>Project location</td><td></td></tr>
<tr><td colspan="2">Item</td><td>Result</td><td colspan="2">Index requirement</td><td>Remarks</td></tr>
<tr><td rowspan="6">Mesh</td><td>Type</td><td></td><td colspan="2">Flexible machine-woven wire mesh or flexible plastic mesh</td><td></td></tr>
<tr><td>Wire diameter of flexible machine-woven wire mesh</td><td></td><td colspan="2">≥ 2.0 mm</td><td></td></tr>
<tr><td>Mesh diameter</td><td></td><td colspan="2">50 mm to 75 mm</td><td></td></tr>
<tr><td>Maximum stretching force of flexible plastic mesh</td><td></td><td colspan="2">≥ 6.0 kN/m</td><td></td></tr>
<tr><td>Ageing resistance of flexible plastic mesh</td><td></td><td colspan="2">≥ 15 years</td><td></td></tr>
<tr><td>Corrosion protection</td><td></td><td colspan="2"></td><td></td></tr>
<tr><td colspan="2">Inspector</td><td colspan="4">Signature： Date:</td></tr>
</table>

Appendix D Inspection Contents and Record Formats of Vegetation Concrete

Table D Inspection contents and record formats of vegetation concrete

<table>
<tr><th>Project name</th><th colspan="3"></th><th colspan="3">Project location</th></tr>
<tr><th rowspan="2">Item</th><th colspan="3">Result</th><th colspan="3">Index requirement</th></tr>
<tr><th>1 d</th><th>3 d</th><th>≥28 d</th><th>1 d</th><th>3 d</th><th>≥28 d</th></tr>
<tr><td>Bulk density</td><td colspan="3"></td><td colspan="3">1.3 g/cm^3 to 1.7 g/cm^3</td></tr>
<tr><td>Aeration porosity</td><td colspan="3"></td><td colspan="3">≥25 %</td></tr>
<tr><td>pH</td><td colspan="3"></td><td colspan="3">6.0 to 8.5</td></tr>
<tr><td>Available moisture content</td><td colspan="3"></td><td colspan="3">≥15 %</td></tr>
<tr><td>Hydrolyzable nitrogen</td><td colspan="3"></td><td colspan="3">≥60 mg/kg</td></tr>
<tr><td>Available phosphorus</td><td colspan="3"></td><td colspan="3">≥20 mg/kg</td></tr>
<tr><td>Rapidly available potassium</td><td colspan="3"></td><td colspan="3">$\geq 1.0\times10^2$ mg/kg</td></tr>
<tr><td>Restored degree of shrinkage</td><td colspan="3"></td><td colspan="3">≥90 %</td></tr>
<tr><td>Unconfined compressive strength (MPa)</td><td></td><td></td><td></td><td>0.25 to 0.45</td><td>0.40 to 0.55</td><td>≥0.38</td></tr>
<tr><td>Erosion modulus at 80 mm/h rainfall intensity</td><td></td><td colspan="2"></td><td>$\leq 3.0\times10^2$ $g/(m^2\cdot h)$</td><td colspan="2">$\leq 1.0\times10^2$ $g/(m^2\cdot h)$</td></tr>
<tr><td>Inspector</td><td colspan="6">Signature： Date:</td></tr>
</table>

Explanation of Wording in This Code

1 Words used for different degrees of strictness are explained as follows in order to mark the differences in executing the requirements in this code:

1) Words denoting a very strict or mandatory requirement:

"Must" is used for affirmation; "must not" for negation.

2) Words denoting a strict requirement under normal conditions:

"Shall" is used for affirmation; "shall not" for negation.

3) Words denoting a permission of a slight choice or an indication of the most suitable choice when conditions permit:

"Should" is used for affirmation; "should not" for negation.

4) "May" is used to express the option available, sometimes with the conditional permit.

2 "Shall meet the requirements of..." or "shall comply with..." is used in this code to indicate that it is necessary to comply with the requirements stipulated in other relative standards and codes.

List of Quoted Standards

GB/T 50123,	*Standard for Soil Test Method*
GB 50218,	*Standard for Engineering Classification of Rock Mass*
GB 51018,	*Code for Design of Soil and Water Conservation Engineering*
GB 175,	*Common Portland Cement*
GB 2772,	*Rules for Forest Tree Seed Testing*
GB 5084,	*Standards for Irrigation Water Quality*
GB/T 16453.4,	*Comprehensive Control of Soil and Water Conservation—Technical Specification—Small Engineering of Store, Drainage and Draw Water*
GB/T 20203,	*Technical Specification for Irrigation Projects with Low Pressure Pipe Conveyance*
CJJ 82,	*Code for Construction and Acceptance of Landscaping Engineering*
DL/T 5353,	*Design Specification for Slope of Hydropower and Water Conservancy Project*
HJ/T 166,	*The Technical Specification for Soil Environmental Monitoring*
LY/T 1970,	*Organic Media for Greening Use*
SL 419,	*Test Specification of Soil and Water Conservation*